SNEAKY PRESS

©Copyright 2021

Pauline Malkoun

A catalogue record for this work is available from the National Library of Australia.

ISBN 9781922641144

Sneaky Press is the imprint of Sneaky Universe.
www.sneakyuniverse.com
First published in 2021

Sneaky Press
Melbourne, Australia.

The Book
of
Random Sleep Facts

Sneaky Press

Contents

Facts about why we sleep

We all need to sleep, (eventually our bodies will shut down and sleep whether we want to or not) but researchers are still not 100% sure why. There are 2 main theories — restoration theory and evolutionary theory.

Restoration theory suggests that sleep provides time to help us recover from activities during waking time that use up the body's physical and mental resources.

Restoration theory suggests that NREM and REM sleep tend to have different restorative effects.

NREM sleep is considered to be important for restoring and repairing the body including physical growth, tissue repair and recovery especially during NREM stages 3 and 4 of NREM, when the brain is least active.

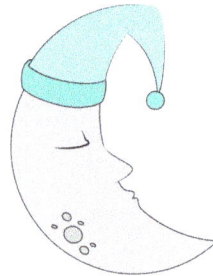

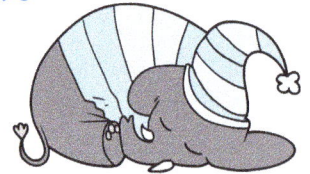

It is thought that REM sleep may assist in the forming new memories.

Evolutionary theory suggests that the reason we sleep is to keep our species from becoming extinct and enhance our survival.

It suggests that sleep evolved to enhance our survival as a species by protecting us because it makes us inactive during the part of the day when it is most dangerous to move around.

According to this theory, once a person (or animal) has had it's survival needs such as eating, drinking, caring for its young and reproducing met, it must spend the rest of its time saving energy, hidden and protected from predators.

While we are sleeping, we are not interacting with the environment and therefore less likely to attract the attention of potential predators and get into dangerous situations.

The Stages of Sleep

During a typical night, we experience two very different types of sleep - NREM sleep (Non-Rapid Eye Movement) and REM sleep (Rapid Eye Movement) .

There are 4 stages of NREM sleep.

We spend about 3/4 of our total sleep time in NREM sleep.

It takes about 45 to 60 minutes to progress through the first NREM sleep cycle from stage 1 to stage 4 before we gradually move back up through stages 3 and 2 to REM sleep.

Each stage of sleep has a distinguishable pattern of brain wave activity.

The average length of a complete NREM–REM sleep cycle is about 90 minutes.

As the night goes on, we have more REM sleep.

Facts about NREM Stage 1

Most people enter sleep through NREM

The point at which we fall asleep is called Sleep Onset.

NREM Stage 1 is indicated by the body through a decrease in heart rate, breathing, body temperature and muscles start to relax.

As we fall asleep we gradually lose awareness of ourselves and our surroundings.

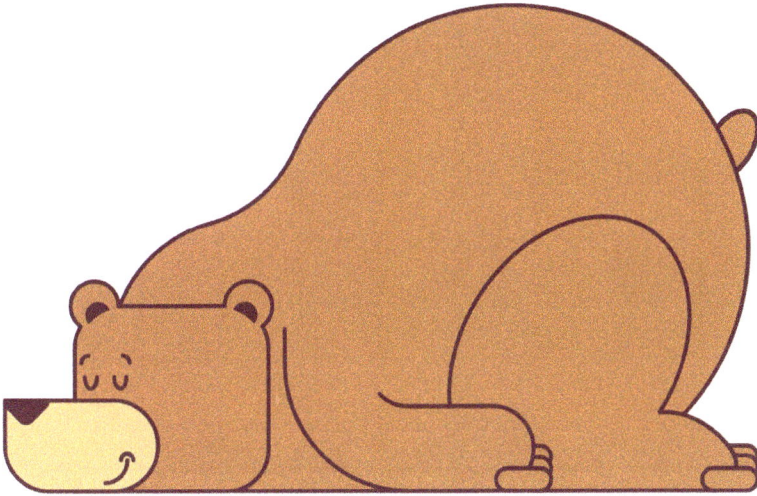

NREM Stage 1 accounts for about 4 or 5% of the total sleep time.

We can be easily awakened during stage 1 by sound and touch, for example, a ringing phone or feeling a blanket being draped over the body.

If awoken during stage 1, we may feel as if we haven't been asleep at all.

Facts about NREM Stage 2

NREM Stage 2 is the point when people are considered to be truly asleep.

NREM stage 2 is light sleep, so a sleeper in stage 2 is less easily disturbed than it is in stage 1. The phone needs to ring loudly or a door needs to be slammed to wake someone from this stage.

If woken from the first half of this stage, most people report that they really didn't think they were asleep, but were just dozing or thinking.

We spend about half our total sleep time each night in Stage 2 REM sleep.

About halfway in NREM Stage 2, people are unlikely to respond to anything except extremely strong or loud noise or touch—maybe being shaken could do the job!

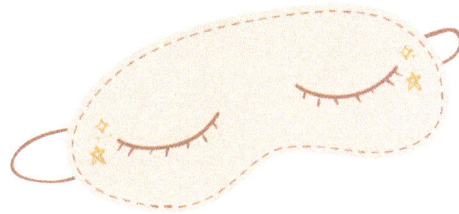

The first time a sleeper gets to Stage 2, they will spend between 10 and 25 minutes. This lengthens with each successive cycle.

Facts about NREM Stage 3

NREM Stage 3 is considered the start of the deep sleep.

We spend less than 10% of our total sleep time in NREM Stage 3.

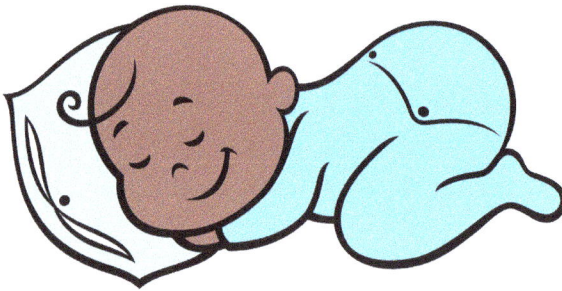

There may not be any stage 3 NREM sleep during the second half of the night.

When in NREM Stage 3, we are extremely relaxed, and become even less likely to respond to noise.

It is difficult to wake someone from NREM stage 3. If they are woken, they are disorientated unable to think clearly at first.

Facts about NREM Stage 4

Stage 4 is the deepest stage of sleep.

In NREM Stage 4, our body is completely relaxed and we barely move. Heart rate, blood pressure and body temperature are at their lowest.

It is very difficult to wake someone from NREM Stage 4.

If someone is woken from NREM Stage 4, they will need a few minutes to orient themselves.

As the night goes on, time spent in NREM stage 4 decreases and even stop occurring.

A person may spend between 20 to 40 minutes in NREM stage 4 in the first sleep cycle.

Overall, we spend about 10-15 % of our sleep time in NREM Stage 4 on a typical night.

Facts about REM Sleep

We spend approximately 20–25% of our total sleep time in REM

The first REM stage that occurs may only last 1 to 5 minutes, the second about 12–15 minutes, the third about 20–25 minutes and so on.

During REM sleep, the brain wave pattern is like that produced during alert wakefulness, but the sleeper looks completely relaxed.

Most dreaming occurs during REM sleep.

Most people dream a few times a night, even if they cannot remember their dreams.

As the night goes on, time in REM sleep increases and gets closer together.

REM sleep is characterised by spontaneous bursts of rapid eye movement during which the eyeballs quickly move under the closed eyelids, zooming back and forth and up and down.

Random facts about Dreams

We dream during both REM and non-REM sleep.

The dreams we have during REM sleep are usually weirder than the ones we have during non-REM sleep, which tend to be repetitive.

Today, about 10% of people dream in black and white – the rest of us dream in colour. Before colour television, only 15% of people dreamed in colour.

Women dream about men and women equally, while men dream about other men 70% of the time.

Dreams are difficult to remember. You forget about half within 5 minutes of waking up and forget about 90% after another 5 minutes.

Humans spend about 6 years of their life dreaming.

You cannot dream about faces you haven't already seen.

Random facts about Sleep Disorders

There are more than 80 different sleep disorders divided into 2 main types.

Parasomnias include disruptions to sleep as a result of an abnormal sleep-related event such as sleep walking, teeth grinding or terrifying dreams.

Dyssomnias include problems with the sleep-wake cycle such as having trouble falling or staying asleep, not being able to stay awake or sleeping at the wrong times.

Insomnia is the most common sleep disorder — with an estimated 30% of adults having symptoms of insomnia at some time in their life.

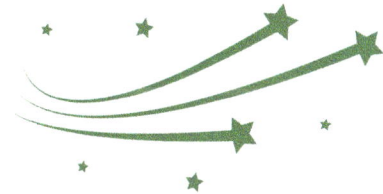

5–10% of adults have a insomnia over a long period of time.

Random facts about Sleep Walking

The scientific name for sleep walking is somnambulism.

Sleep walking involves getting up from bed while still asleep and walking around and can include performing other behaviours like getting dressed.

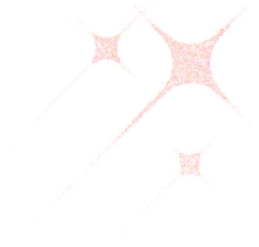

It's thought that up to 15% of the population are sleepwalkers.

A sleep walker will usually return to bed, lie down and continue to sleep without awakening if left alone.

Sleep walking episodes may occur up to 3 or 4 times a week.

Sleep walking is very common in children. It is thought that between 10–30% of children have had at least one sleep walking episode, and that 2 – 3% sleep walk often.

Generally, a sleep walking episode only lasts for a few minutes and is rarely more than 15 minutes, but sleep walking that lasted as long as an hour has been recorded.

Sleep walking usually occurs during the deep sleep of Stages 3 and 4 NREM.

Random facts about Sleep Deprivation

When we do not get enough sleep, we experience sleep deprivation.

Partial sleep deprivation occurs when we get less sleep than what is normally required.

Total sleep deprivation occurs when we get no sleep at all over a short or long-period of time.

Sleep deprivation impacts our ability to process our own emotions, understand the emotions of others and manage our emotional reactions.

The record for the longest time anyone has gone without sleep is 18.7 days.

When people are sleep deprived, they may slip into a microsleep. A microsleep is a very brief period of sleep that lasts for up to a few seconds while a person is awake.

Sleep deprivation impacts our ability to pay attention.

Sleep deprivation can impact our ability to control our behaviour, for example being naughty or making silly choices.

Sleep deprivation has been known to be linked to higher physical injury rates.

Sleep deprivation negatively impacts thinking speed and accuracy.

Facts to Help You Sleep Better

Having a regular sleep schedule — that is waking up and going to bed at the same time every day (including the weekend) will help.

Avoiding unpleasant activities, conversations and thinking about problems just before bedtime will help you sleep better.

Make sure you get enough natural light during the day—it helps maintain your sleep-wake cycle and therefore you will sleep better.

Doing exercise during the day — preferably in the morning or at least 4 hours before bedtime can help you sleep better.

Do not do any activities that cause a lot of excitement or moving around too much (this includes exercise and playing video games) — these will not help you sleep — in fact, they will wake you up and make sleep onset difficult.

Having a nap that is longer than 30 minutes or very close to bed time will not help you sleep better.

When you cannot sleep you should get up out of bed and go do something else.

Random facts about Sleep

After the birth of a child, parents lose between 400 and 750 hours of sleep in the first year.

People slept an average of 9-10 hours per night before the invention of electricity.

These days, 30% of adults sleep less than 7 hours per night.

People are more likely to nod off at 2 am and 2 pm more easily than other times.

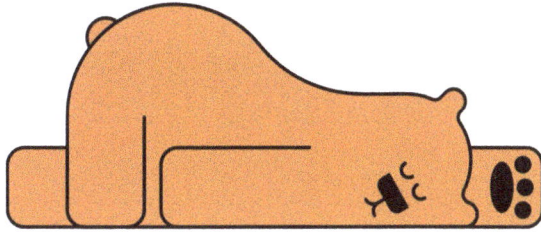

Almost everything we know about sleep was discovered in the last 50 years.

We can only snore during NREM sleep.

Adults who regularly get less than 7 hours of sleep per night are more likely to get sick than those who sleep more than 7 hours each night.

Sleep requirements change with age.

From birth, the total amount of time we spend sleeping gradually decreases as we get older.

More random facts about Sleep

According to NASA (yes the space people), the perfect nap lasts for exactly 26 minutes

People can not sneeze while sleeping—it is impossible.

Research has found that counting sheep is not an effective way of inviting sleep onset. It seems to be too boring; imagining a calm landscape generally works better.

Most people burn less calories while they watch TV than when they are asleep.

Using electronic devices in the two hours before bed time can affect your sleep. They emit blue light which tricks your brain into thinking it's daytime.

Humans are the only mammals that willingly delay sleep.

While asleep, the brain selectively filters out noises that might wake you up as you sleep – especially noises that don't suggest you're in danger.

Even more random facts about Sleep

It takes 7 minutes for the average person to fall asleep.

Each additional child in a household increases a mother's risk of becoming sleep deprived by 46%.

People who lose their ability to see later in life can still see in their dreams.

To protect their fancy hairstyles, rich ancient Egyptians slept with uncomfortable neck supports instead of pillows.

Before alarm clocks were invented, factories employed people to knock on the bedroom windows of their workers with a long stick, to ensure they arrived at work on time.

Somniphobia is the fear of falling asleep.

Oneirophobia is the fear of nightmares or dreams.

Clinomania is the irresistible urge to stay cozy in bed all day, while dysania is the word for that feeling when you've just woken up and really don't want to get out of bed.

Random facts about Animal Sleep

Koalas can sleep 18-20 hours every day

Snails can sleep for three years at a time.

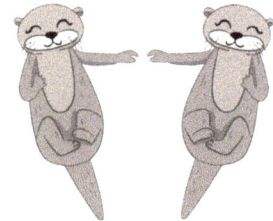

Giraffes can get by on an average of less than 2 hours sleep a night.

Sea otters sleep holding hands so they don't drift away from each other.

Cows and other hooved animals sleep standing.

Sloths and bats sleep hanging upside down.

Nocturnal animals such as possums and wombats sleep during the day.

When whales and dolphins sleep, only half their brain rests at a time so that they can come up for air.

Cats sleep for 70% of their lives.

z z z

Sleep idioms

To sleep like a log is to sleep very deeply — if someone is sleeping like a log, they are very difficult to wake — Which stage of sleep do you think they are in? (Check your answer on p.16)

A catnap is a short sleep in the day.

To sleep on it means to think about something overnight before making a decision.

To lose sleep over something is to worry about it.

To let sleeping dogs lie is to avoid interfering with a situation to make sure you do not cause trouble.

If someone is described as a night owl, they prefer to stay up and work late.

Sleep idioms

To get up on the wrong side of the bed means to start the day in a bad mood that lasts all day.

To go out like a light means to fall asleep very quickly.

A beauty sleep is a stretch of sleep which will keep one young and beautiful—old people need lots of beauty sleep!

To nod off is to fall asleep without meaning to, usually while you are supposed to be awake. We are more likely to nod off if we find something boring.

To not sleep a wink means to not get any sleep at all—what else do we call this? (see p.28 to check your answer)

To burn the candle at both ends is to go to sleep late and wake up early.

Sleep Jokes

Why do dragons sleep during the day ?

So they can fight knights!

Where do fish sleep?

On the river bed!

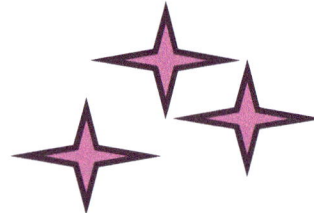

What kind of stories does a Mummy cow read to her baby cow before bed?

Dairy tales

Did you hear about the kidnapping earlier today?

It's ok, he woke up.

What do you call a
sleeping dinosaur?
A dino-snore!

Which animals
sleep with their
shoes on?

Horses

Did you hear about the
man who was dreaming
that he was eating a
giant marshmallow?

He woke up and his
pillow had disappeared.

Why did the meatball
sauce tell the
spaghetti to close its
eyes and go to sleep?

It was pasta
bedtime!

z.z.z

www.ingramcontent.com/pod-product-compliance
Lightning Source LLC
Chambersburg PA
CBHW042334030426
42335CB00027B/3334